These students from Ms. Tiber's class felt very sad. The park near their school was not open anymore.

"Where can we play now?" asked Tracy.

Ms. Tiber could see the sadness in their faces. She said, "We can work to keep the park open. You can write letters to people in our local government."

Tracy and Roland made large posters. Macy and Ira wrote a letter to the school paper. Selby Anne spoke at a school meeting.

"People will see each poster," said Ms. Tiber. "Then they will want to help too."

At school the next day, Selby Anne said, "We can fix up the park! We know how to be helpful."

Ms. Tiber started a list of items they would need. The students added even more items to the list.

That weekend everyone worked at the park. The students were careful as they painted. Soon every park bench looked bright and fresh.

Later the students took armfuls of plants to the park. People helped them dig holes. They said, "We are very grateful to you."

That afternoon, Mr. Egan spoke to the students.

"The local government will keep this park open," he said. Then he gave each student a big badge that said Friend of the Park.

The End